MEMORIAL TO THE LEGISLATURE
OF THE STATE OF NEW YORK

FOR AN INVESTIGATION OF THE CONDITIONS SURROUNDING GAS AND ELECTRIC LIGHTING IN THE CITY OF NEW YORK

THE MERCHANTS' ASSOCIATION OF NEW YORK

January 11, 1905

1837
ARTES
SCIENTIA
VERITAS
LIBRARY OF THE
UNIVERSITY OF MICHIGAN
TUEBOR
SI·QUÆRIS·PENINSULAM·AMŒNAM
CIRCUMSPICE

MEMORIAL

For an Investigation of the Conditions Surrounding Gas and Electric Lighting in the City of New York.

To the Honorable the Legislature of the State of New York:

THE Merchants' Association of the City of New York presents this memorial and petition to the Senate and the Assembly of the State of New York, and begs the passage of a resolution authorizing the appointment of a legislative committee empowered to examine into, report upon and make recommendations concerning:

First—The conditions surrounding the production, distribution and cost of gas and electric light in the City of New York and the reasonableness of the charges therefor, both to the City and to private consumers.

Second—The conditions surrounding the operation and control of the electrical subways constructed in the City of New York under the provisions of the statutes of this State; the accounts and earnings of such subways and the amounts, if any, due the City as the result of their operation.

FIRST.

AS to the conditions surrounding the production, distribution and cost of gas and electric light in the City of New York, and the reasonableness of the charges therefor, both to the city and to private consumers, it is respectfully urged:

These conditions and the question of the fairness of the price which both the city and private consumers pay for gas and electric light need thorough investigation. The City of New York is compelled to purchase lights for its streets, public parks and public buildings from the various gas and electric companies, doing business in the city. Under the provisions of the city charter it is made the duty of the Commissioner of Water Supply, Gas and Electricity of the City of New York to advertise for bids for public lighting and to execute contracts therefor after such public bidding, the term of such contracts not to exceed one year. The conditions which have infallibly been found to prevail as a result of the effort of the said city department to obtain contracts for lighting are set forth in reports made by the department to the Mayor and to the Board of Estimate and Apportionment of the city during the years 1902 and 1903. When the city administration elected in the fall of 1901 came into office in January, 1902, the contracts for public lighting had substantially been awarded, and there was no opportunity for an inquiry into the reasonableness of the charges asked by the gas and electric lighting companies for public service, until the year 1903; but the Commissioner in charge of the department in the early part of 1902 reported to the Mayor of the City on March 31, 1902, as follows:

> "The City of New York pays to the gas and electric light companies, for lighting its streets, highways, parks, other public places and public buildings, more than three million dollars per annum. My estimate of March last made the figures $2,914,041.48, but this estimate, as was afterward ascertained, did not include numerous contracts which it has heretofore been the custom of the various city departments to make, independently of the Department of Public Buildings, Lighting and Supplies. The amount of these contracts has not been as yet fully determined, but it probably aggregates $225,000. Unfortunately, under existing business conditions, the City is subject to monopoly rates in awarding these contracts, for there is no genuine competition for bids. The New York Edison Company, the Consolidated Gas Company, the Brooklyn Edison Electric Illuminating Company, not to speak of other combinations, are without rivals in their particular field, and the City is forced to pay whatever price these corporations decide to

charge. Whether, under such circumstances, municipalization of gas and electric light would not be an advantage, is an inquiry worth profound consideration. If the City's gas and electric light business were conducted with due regard to the City's interests, there would, I think, be but one answer to the question."

On March 12, 1903, the Honorable Robert Grier Monroe, then Commissioner of Water Supply, Gas and Electricity, transmitted to the Board of Estimate and Apportionment a report stating that the bids for gas and electric lighting for the year 1903 which had been submitted to him, should one and all be rejected under the power conferred upon him by the city charter, on the ground that, within five years prior thereto, the lighting interests in the city had all practically been united, with the result of preventing all competition between rival companies operating in the same field, and that even in boroughs where there had been no formal consolidation, the territory had been so apportioned between the bidding companies as to preclude all competition between them. The report further set forth that the Consolidated Gas Company of New York controlled all the gas and electric light facilities in the borough of Manhattan as well as all gas and electric light facilities in the more important sections of the borough of The Bronx, with the effect, so far as gaslight was concerned, of not only keeping prices at a maximum but retarding all improvement in the utilization of gas. The report further stated:

"The appropriation for all public lighting for this year, 1903, is $3,306,346.23. The corresponding appropriation in 1898, the first year of Greater New York's existence, amounted to $2,570,001.88. The increase in five years has been $736,344.35, or between 28 and 29 per cent. Notwithstanding this increase, the sum appropriated this year, in view of the prices asked, is entirely inadequate to meet the reasonable requirements of the City. There can be little improvement in the densely populated sections which are now insufficiently lighted, and large areas which have recently been improved must, for the time being, be left without any lights at all.

"The growth of the City, the many public improvements, which are now under way—as, for example, the new bridges which are rapidly nearing completion—all mean a great increase of public lighting; and unless there is a material reduction in prices the annual appropriation five years hence must necessarily equal or exceed $5,000,000."

The report further showed that, in the borough of Manhattan, the price for a 2,000-candle-power electric lamp was $146; in Brooklyn, for a 1,200-candle-power lamp, $124.50; whereas according to reliable *data,* obtained by the department, the prices charged by various electric light companies, to sixty-eight other cities where conditions similar to those in New York exist, for supplying such cities with 2,000-candle-power lamps, averaged $88.60, and to twenty-three other cities similarly situated using 1,200-candle-power lamps, averaged $81.08.

The capitalization of the various electric light companies operating within the territory of Manhattan and The Bronx, all under the control of the Consolidated Gas Company of the City of New York, was shown in the report to aggregate, for stock and bonds, about $30,000,000, a sum far in excess of the actual value of the several plants and properties, while the monster consolidation formed by their union was capitalized at $85,538,000; $45,200,000 in stock and a bonded debt of $40,138,000, this enormous valuation being based upon the profit permitted by the monopoly system in existence. It was further shown that modern improvements, especially in the concentration of the work of producing power, had, in recent years, effected a decrease in the cost to the companies of the light itself, reducing the cost to the Consolidated Gas Company to about three cents per kilowatt hour of current delivered; but, notwithstanding this large recent reduction in the cost of manufacture, there had been no corresponding reduction in prices to the city or to the consumers, in fact, no reduction at all, the bills presented to the Department for lighting public buildings for January, 1903, showing that the city was charged upon an average at the rate of ten and fifteen-one-hundredth cents per kilowatt hour, or nearly three hundred and fifty per cent. of cost.

The report fairly proved that the boroughs of Manhattan and

The Bronx of the City of New York were at the mercy of a lighting combination, able to fix prices for public lights without any competition whatever; that the companies in control were immensely overcapitalized; that their prices were fixed to meet dividends and interest charged on excessive overcapitalization, and that the prices charged included excessive profits far beyond the cost of manufacture and delivery. It concluded by recommending that the city be empowered to establish and maintain an electric plant for lighting its streets, parks and other public places. In the report were also set forth figures showing the excessive charges made for gas lighting and the inability of the department to compel the companies furnishing gas to fulfil the requirements of purity and illumination fixed by law.

The borough of Brooklyn is equally under the subjugation of a gas and electric light monopoly. The electric light and power companies are owned or dominated by the Kings County Electric Light and Power Company. All rival and competing gas-producing companies have within a few years been merged into one larger aggregate known as the Brooklyn Union Gas Company, which is immensely overcapitalized, and which, in order to pay dividends upon its inflated capital and heavy interest charges, exacts from the city and private consumers unnecessarily high rates and at the same time furnishes an exceedingly poor illuminant. The story of its inflated capitalization and the methods devised to consummate the merger, is told in a recent suit between the promoters of the deal. Whether the water gas, so-called, which is the product of the Brooklyn Union Gas Company, is dangerous to life and health, is a question requiring expert consideration and decision. It is, however, often alleged, and with some show of reason, that this gas is deleterious in its effects upon the human system. Complaints of its insufficiency as an illuminant are made all over the borough. One has but to be a resident in that borough to be made aware of the fact that either the standard of illumination fixed by law is so criminally inadequate as to imperil eyesight, or that the light which is furnished is far below that standard. The effect of this inadequate illuminant upon the eyes of the people, however much light is consumed and however large the consumers' bills, is pernicious in the extreme. The light appears to decrease in an inverse ratio to the size of the combination and the amount of its profits.

In consequence of the report of Commissioner Monroe to the Board of Estimate and Apportionment advocating the passage of legislation to enable the city to establish and maintain an electric lighting plant of its own, a bill was introduced in the Legislature early in 1903 and its passage urged by the municipal authorities and by numerous civic organizations, of which this Association was one.

One unanswerable argument, as we deemed it, in favor of the bill, was, that if the City of New York should undertake to supply itself with electric light, it would have various decided advantages over the consolidation beneath whose yoke it remains, for it would be subject merely to low-interest payments, based on loans measured by the funds actually required for the construction of a plant, but would be emancipated from paying the fixed charges which must accompany inflated capitalization. In the borough of Manhattan, where electrical energy is almost universally transmitted through subways, the city would be free of any expense by way of rental for the space occupied in the subway, because the acts of the Legislature under which the electric subways have been constructed, and the contracts made with the subway companies specifically provide that the city shall, without charge, be furnished with all spaces in said subways necessary for its electrical conductors, and the electrical conductors of each separate department of the city; and that if accommodation is not practicable in existing subways, more must be built to meet the city's needs.

The bill to authorize the city to establish a municipal lighting plant was never favorably reported upon by either the Senate or the Assembly. Consequently that measure of relief was denied, which, in the judgment of the then municipal authorities and of the various civic organizations supporting the bill, its passage would have given.

We are glad to observe that in his message of January 1, 1905, to the Board of Aldermen of the City of New York, the Honorable Mayor of the City advocates the enactment of legislation to enable the City to erect and maintain an electric plant "to light the streets, parks and public buildings of the City," and has caused a bill for that purpose to be prepared for introduction into the Legislature. In his message the Mayor also states: "The prices which the City is compelled to pay for gas and electric light are so out of proportion with the charges in other cities, that they must be extortionate."

The Legislature in 1903 having seen fit to refuse to empower the City to establish an electric-lighting plant for city lighting, the City has continued to remain under the control and at the mercy of the several lighting companies, and private consumers are still forced to pay the exorbitant charges exacted, and for insufficient light. If the Legislature is not willing to authorize the City to establish an electric-lighting plant of its own, it should at least investigate this entire subject and by its investigation either confirm or refute the widely prevalent belief that the prices charged both the City and private consumers are far too high and that the illuminants are inferior in quality. Even if the City should be empowered to construct an electric-lighting plant (and we believe that should be done), inasmuch as that plant would not carry relief to private consumers, the investigation sought is eminently necessary and proper. Some idea of the burden placed upon private consumers of gas is given in an official report of the U. S. Government, from which we cite below.

In the Fourteenth Annual Report of the Commissioner of Labor of the United States for 1899, upon "Water, Gas and Electric Plants and Private and Municipal Ownership," there appears an exhaustive analysis of the cost of gas production in more than one-half of all the gas plants of the country, from which the deduction is made by the Commissioner that under the most favorable and most economical conditions of production, in plants of large capacity, the average cost of gas does not exceed 42 cents per one thousand cubic feet of gas sold. The cost as thus ascertained includes not only all the elements of manufacturing cost, but also allowance for waste or leakage, depreciation, interest on investment and taxes.

In the cases cited the Commissioner says: "The total income for the year was $3,405,781, and the cost of production, including taxes, depreciation and interest on investment at the average rate paid on the last issues of bonds in the cities involved (3.5 per cent.), was $1,657,845, leaving a net profit of $1,747,936, over and above the cost," or a profit in excess of 100 per cent. in addition to 3½ per cent. interest.

Even though the City be empowered to erect its own plant, under existing monopoly conditions, the purses of its citizens would still continue to be drained to pay these enormous profits.

The urgency and importance of an investigation at this juncture into the charges for public lighting are emphasized by the fact that, within a few weeks past, contracts have been made with the gas and electric-lighting combine for the supply of these illuminants to the City's streets, buildings and parks, at the extortionate prices of which the late Commissioner of Water Supply, Gas and Electricity complained, prices shown by him to be far in excess of the rates charged by corresponding light-producing companies to other cities of the United States. According to the information compiled by the Department of Water Supply, Gas and Electricity, the City of New York pays the electric-lighting monopoly a price for lights which represents a profit of 250 per cent. The profit from the sale of gas, if not quite so great, far exceeds a reasonable percentage. In fact, the excessive nature of these charges is confirmed by a subcontract between the City and these companies, by which the companies agree, in certain cases, to make slight deductions from their prices; but even with these deductions, the charges to the City are altogether exorbitant and unreasonable. That these rates are excessive and unjust appears to have been recently held by one of the Justices of the Supreme Court of this State, in an action instituted by a tax-payer of the City to enjoin the City officials from carrying these contracts into effect. The press of the City, almost without a single dissenting voice and irrespective of party considerations, has united in denouncing the execution of the contracts and in commending the action of the tax-payer.

SECOND.

NOR would the investigation upon which we respectfully beg the Legislature to enter, be complete, unless it also took into consideration the conditions surrounding the operation and control of the electrical subways in the City and the inquiry reached so far as to ascertain their earnings and what amounts, if any, are due from them to the City of New York, as the result of their operation.

The overhead-wire conditions in the former City of New York (the present Borough of Manhattan), became, years ago, so intolerable that the Legislature, in 1884, directed that all electrical conductors be placed underground; and, in 1885, it passed an act establishing in cities of large population, a board of Commissioners of Electrical Subways whose business it should be to devise a plan and scheme for the construction of electrical subways suitable for the electrical conductors of the companies using aerial wires in their business, the compulsory removal of overhead electrical conductors, and their replacement by wires in underground conduits. Subway companies were organized, and the construction of subways and the leasing of subway space, under the supervision of the City authorities, were delegated to such corporations; and the law of 1885 was amended to permit such delegation. Under this legislation, the Consolidated Telegraph and Electrical Subway Company was formed and an agreement was made between it and the City of New York, by which it was to build the subways and lease space therein. When it was discovered to be dangerous to operate high and low tension wires in the same conduits, a new agreement was made with the Consolidated Company for the construction of separate subways for conductors carrying a high potential. The Empire City Subway Company, Limited, was incorporated to build subways, conduits and ducts for telegraph and telephone wires and for the low-tension electric light and power conductors of the Edison Electric Illuminating Company of New York (one of the companies since absorbed by the Consolidated Gas Company). The Consolidated Telegraph and Electrical Subway Company is said to be owned by the New York Edison Company, and this last-named company is the property of the Consolidated Gas Company. Thus both subway companies appear to be under one control—that of the lighting monopoly.

Under the aforesaid legislation and the contracts entered into between the City, through the Board of Commissioners and the Consolidated Telegraph and Electrical Subway Company, and the Empire City Subway Company, Limited, the City reserves the right, without charge, to all subway space necessary for its electrical conductors and the electrical conductors of each of its separate departments, and may compel the companies to furnish it with additional space, if necessary, by the construction of new subways; but it is credibly asserted that the Consolidated Telegraph and Electrical Subway Company has refused to accede to the demands of the City for requisite space and that, because of its refusal, some of the City departments, notably the Fire Department, has been obliged to maintain overhead wires within the Borough of Manhattan. The contracts and legislation further provide that whenever the net annual profits of each company, after the deduction of its operating expenses, shall exceed 10 per cent. upon the actual cash capital investment, the excess of such profit over 10 per cent. shall be paid into the treasury of the City of New York; and each Company is also obligated to keep just, true and full books of account, showing in detail all its transactions, including the amount of subway space leased to various electrical companies, with the number and kind of electrical conductors therein, and the rentals in each instance, such books to be at all times accessible to the Comptroller of the City of New York. Notwithstanding these provisions, the Consolidated Telegraph and Electrical Subway Company has failed to pay to the City such excess of net profit over 10 per cent. and an action was recently instituted by the City of New York to compel an accounting and to recover the share of profit to which the City is entitled. A similar suit is pending against the Empire City Subway Company, Limited. The legislation and contracts prescribe also that at any time after January 1, 1897, upon demand of the Commissioners of the Sinking Fund of the City of New York, all the property of either subway company with all its leases may be transferred to the City upon payment by the City to the company of not less than the actual cost of construction except where the company shall not have earned 10 per cent. on such cost, in which case an additional payment is to be made. The cost of these subways was stated in the Franchise Tax proceedings before the late Judge Robert Earl, to be $6,000,000, but the testimony in the proceeding showed, and Judge Earl found in his decision, that they could be duplicated for $2,650,000. In January, 1904, the Treasurer of the Consolidated Telegraph and Electrical Subway Company filed a sworn statement with the Comptroller to the effect that the cost of construction of electric light and power subways to January 1, 1903, amounted to $7,492,200.62. The Commissioners of Accounts, after an examination of the books of the company in September last (1903), reported that the amount given by the Treasurer of the company is over $3,000,000 above the true cost.

It is the common belief, deduced from the practical identity of the stock interests, that the subway companies are, in some occult but effective manner, articulated with the gas and electric-lighting combine within the former City of New York and that such combination owns or in effect controls the subway companies and the subway space, all of which efficiently works to preclude all possible competition in gas and electric lighting, because if new lighting companies were to be organized it would not be possible for them to conduct their business within the territory of the former city unless their wires were placed in underground conduits; whereas all such subways are under the control of the present lighting monopoly, which, upon various pretexts, would refuse admission into the subways to any probable competitor. Again, under the Transportation Corporations Law, no lighting company can enter the field without the consent of the "municipal authorities," which has been interpreted by the Court of Appeals to mean, the Board of Aldermen, in the City of New York. The present combination has, through its domination of existing companies, obtained the ownership of practically all extant franchise (many, perhaps, of doubtful legality) and if it be as influential a some fear, the difficulty of a competitor's obtaining a nev franchise from the "City authorities" becomes almost insurmount able. Hence by every avenue of approach, threatened competitio is vigilantly averted; it is prevented, firstly, by the consolidatio of the lighting interests, secondly, by the control this consolidatio has gained over the subway companies, and, thirdly, by its owner ship of extant lighting franchises and the difficulty, perhaps im possibility, that would confront any new rival in obtaining a franchise from the Board of Aldermen.

Early in the year 1902, the Commissioner then in charge o the Department of Water Supply, Gas and Electricity, institutec an effort to accomplish the removal of dangerous overheac electric-light wires and their replacement in underground subway within the Borough of Brooklyn. Under his initiative energeti measures were taken toward the consummation of this much to-be-desired end, and the work has since been continued. The conditions in Brooklyn, as described in a report made by the Commissioner to the Honorable Seth Low, Mayor, and Chairman of the Board of Estimate and Apportionment, on June 9, 1902 exemplify the danger to life and property from the continuanc of the present overhead-wire conditions in that Borough, and this danger is intensified by the presence overhead of the heavy trolley feed wires of the several trolley companies.

Concerning the electric light wires the said Commissioner observed in his report:

> "These are a great source of danger in a variety of ways. Strung, as the wires repeatedly are, over or under telegraph and telephone wires, on poles of great height, they are in constant danger of being thrown either to the street or into contact with the low-tension wires, and the moment such contact occurs and the insulation of the wire is sufficiently abraded, the low-tension wire becomes a conductor of the high-tension current. Nor is the danger obvious, because the current may be carried miles, and either a pedestrian upon the street or a lineman attempting to repair, at a distance from the point of contact of the wires, receive a shock

sufficient to kill or permanently injure, or a current might be carried over low-potential conductors into a building, which, not having been safeguarded against the destructive current of a high-potential system, would probably be set on fire, as happened in Brooklyn in February last (1902) at No. 504 Hamilton avenue, office of Nelson Bros., according to a report of the New York Board of Fire Underwriters to me.

* * * * * * *

"During the storms of last February it is a well-known fact that poles gave way or wires were unfastened and thrown in a confused and dangerous network upon the streets, that horses were killed, and, in many instances, human life was endangered. Similar, if not worse, accidents have occurred in other boroughs of the City where wires of varying potential are suspended overhead."

We beg to remind you that in the year 1885 the Senate of this State authorized a full inquiry into the conditions surrounding the cost and distribution of gas in the former City of New York, which elicited a mass of information revealing the utter absence of competition among the ostensibly competing gas companies of that day. The Senate Committee, under the chairmanship of the Hon. Edward B. Thomas (now United States Circuit Judge in the Eastern District of New York), reported that the New York Gas Light Company, the Manhattan Gas Light Company, the Harlem Gas Light Company, the Metropolitan Gas Light Company, the Mutual Gas Light Company, the Municipal Gas Light Company, the Knickerbocker Gas Light Company and the Equitable Gas Light Company had one after another been destroyed or welded into the formidable combination known as the Consolidated Gas Company, organized in 1884, for the purpose of absorbing all extant gas companies operating within the former city. Within a few months after it obtained its charter it had swallowed up all these companies, except the Mutual, the charter of which forbade consolidation with other companies, the total capitalization of the new consolidation being **$39,078,000**. This capitalization of **$39,078,000** was declared by the Senate Committee of 1885 to be worth less than **$20,000,000**. Nearly **$8,000,000** had been issued for alleged franchises and rights, which were worthless and extinct. Genuine competition in gas manufacture has, in fact, never been permitted long to exist in any part of the City of New York.

The advent of the electric light, for a brief time, held out the promise of relief. But hope was soon extinguished. Combination among various electric lighting companies was soon concerted, and forced by stock operators, until in the early part of 1899, all the electric lighting companies were absorbed in the New York Gas and Electric Light, Heat and Power Company. Shortly thereafter a larger consolidation was effected between this company, which was a branch of the Consolidated Gas Company, and the Edison Electric Illuminating Company, under the name of the New York Edison Company, with a capitalization, as above stated, of $85,338,000. The electric lighting companies in Brooklyn are in association with the monopoly in Manhattan, and this monopoly either has extended, or is about to extend, itself so as to embrace the other boroughs of the Greater City. The press of the city recently announced the acquisition by the Consolidated Gas Company of the Westchester Lighting Company, which operates in an important section of The Bronx. If a legislative investigation of the gas consolidation was appropriate, in 1885, an inquiry by you into the conditions which have developed from the alliance of all the lighting interests is far more imperatively demanded at this time. The City of New York is under existing conditions called upon heavily to contribute out of its treasury to the payment of the interest on the bonded indebtedness, and dividends upon the stock, of these heavily overcapitalized combinations, and a similar excessive burden is imposed upon the taxpayers throughout this great municipality.

In whatever guise, however masked, the monster monopoly in control of lights and subways in Manhattan has thus far succeeded in frustrating all efforts of the city and of private consumers to escape from its exactions. For franchises of untold value it has paid no taxes to the State; and, on the other hand, it resists the attempts of the municipal authorities to examine the books of its subway operations or to obtain an accounting as to the city's undoubted shares in the profits of such operations.

If we are correctly advised, none of the gas or electric light companies operating within the greater city have yet paid any franchise taxes.

If the conditions herein portrayed exist—if they are even partially true—the people of the City of New York and the city itself, thus preyed upon and distressed, are, upon the plainest principles, entitled to relief from the Legislature of the State. Whether the facts are such as we believe and assert, your investigation will determine.

Very respectfully submitted,

J. Hampden Dougherty,
Chairman,
F. E. Kinsman,
Charles R. Lamb,
W. A. Marble,
Herman A. Metz,

Committee on Gas and Electricity,
The Merchants' Association of New York.

New York, January 11, 1905.

Filmed 1985 Preservation

www.ingramcontent.com/pod-product-compliance
Lightning Source LLC
LaVergne TN
LVHW020635110826
845149LV00004B/1208

* 9 7 8 1 4 1 8 1 9 1 3 5 1 *